# MY
# SUPERHERO
# FRIENDS

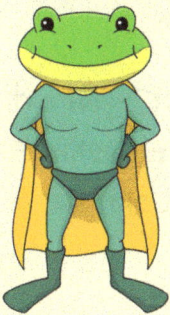

## Math Activity Book

### Tracing
### Counting
### Matching
### and more!

ISBN: 978-1-959247-19-7

THIS BOOK IS DEDICATED TO THE WORLD OF WARCRAFT GUILD
VARIANCE
THANK YOU FOR BEING MY SUPERHERO FRIENDS.
LOVE, KRYSNYA!

Caregivers,

This book aims to introduce math in a fun, easy way so beginners start their math journey confident in their fundamental math skills.

Each section begins with a "What to do." page. We encourage you to teach them how to use it as a reference page when stuck - This also introduces young minds to an additional lifelong skill.

Explain it as a tool: when you need to remember what to do, look at the reference page. Each reference page is the 1st page of that section so they can see the correlation.

It is essential that young learners do this book in order, one page at a time. Each page and section builds on the one before for a smooth transition into the next cognitive connection. Each section starts with 1-4 and then slowly increases until 1-10. This type of slow repetition reinforces understanding along the way.

You can make this book fun; it's an activity book that uses math for its puzzles. Although Dot-To-Dot, Tracing, Coloring, and Matching are all in traditional activity books, this book takes it one step further by using those activities for a fun and enjoyable introduction to fundamental math skills.

Thank you!

# THIS SUPERHERO FRIENDS ACTIVITY BOOK BELONGS TO

This book makes counting and math so much fun. We are so excited you are here to learn with us!

# Section 1

## What to do.

There are lots of activities on page. Take a closer look at each one!

Start from the .1 and follow the dots to make the number.

**Dot to Dot**

Color the number of **Capes**!

Trace each letter slowly with your pencil.

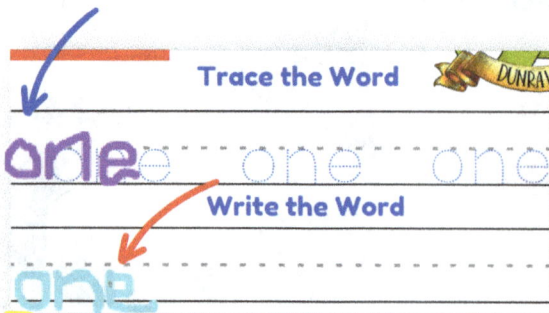

Then you write it on this line. Take your time, try to have the **bottom of the letter on the line** and the **top on the dotted line**!

Find the matching number and color them!

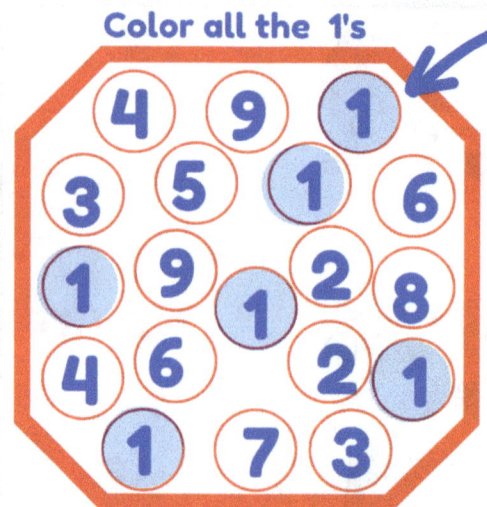

**Color all the 1's**

## Dot to Dot

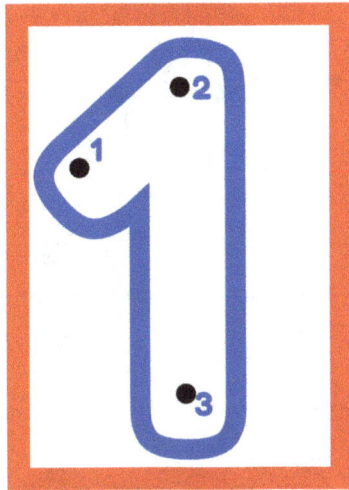

## Trace the Word

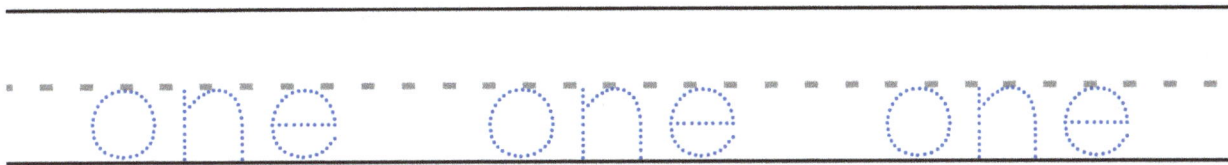

## Write the Word

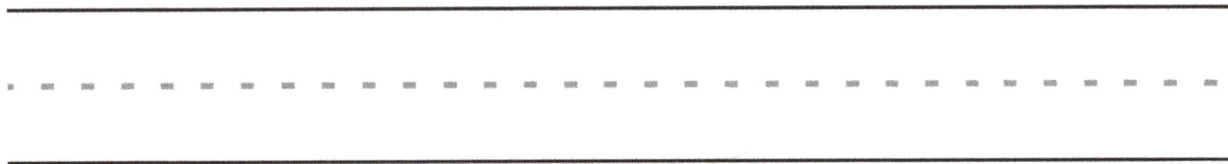

## Color all the 1's

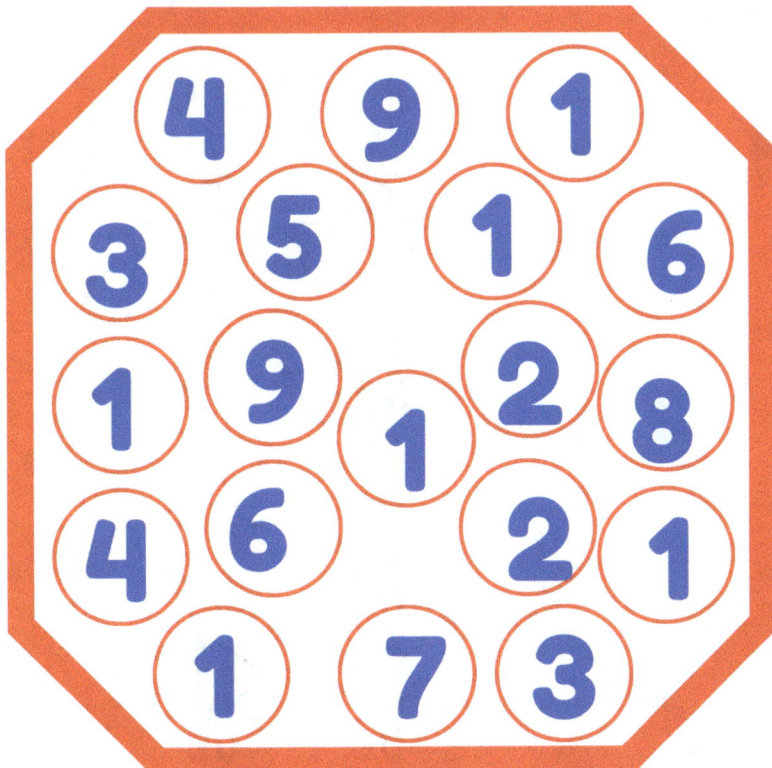

## Color 1 Cape

## Dot to Dot

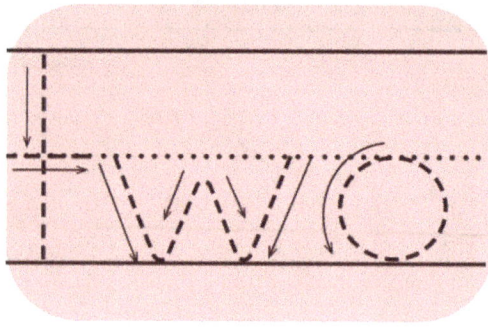

## Trace the Word

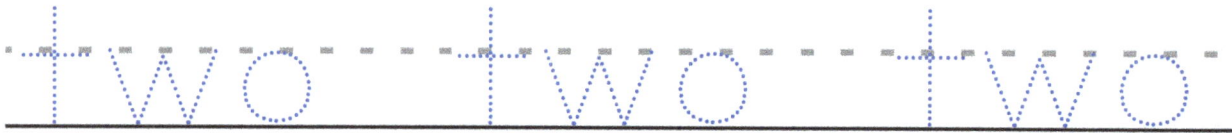

## Write the Word

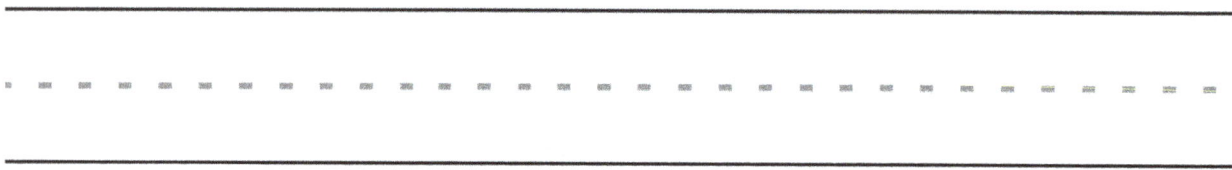

## Color all the 2's

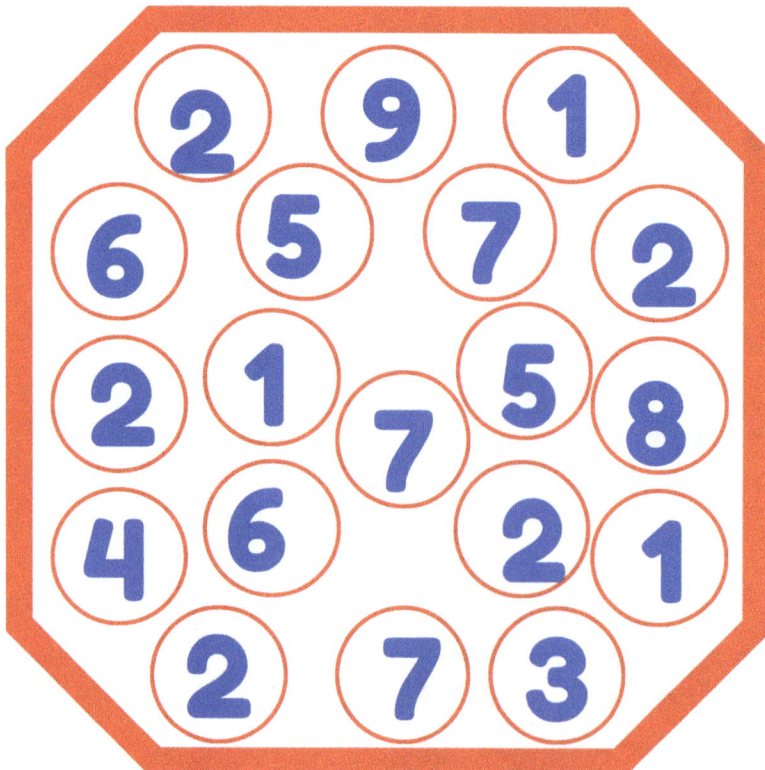

## Color 2 Capes

## Dot to Dot

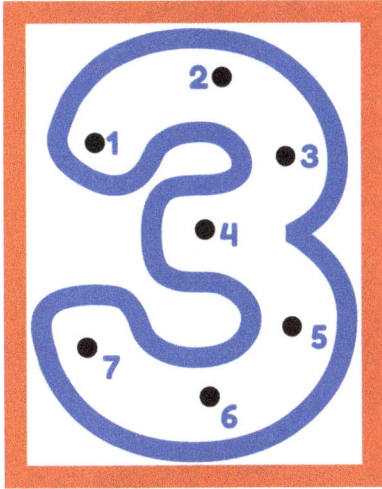

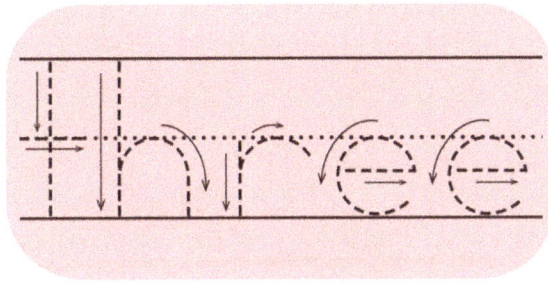

## Trace the Word

three three three

## Write the Word

## Color all the 3's

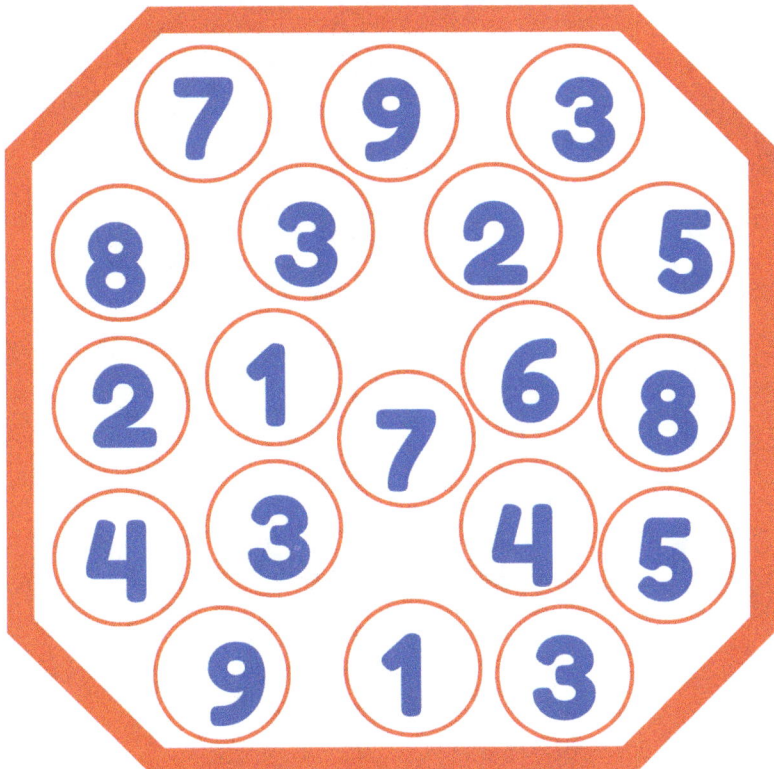

## Color 3 Capes

## Dot to Dot

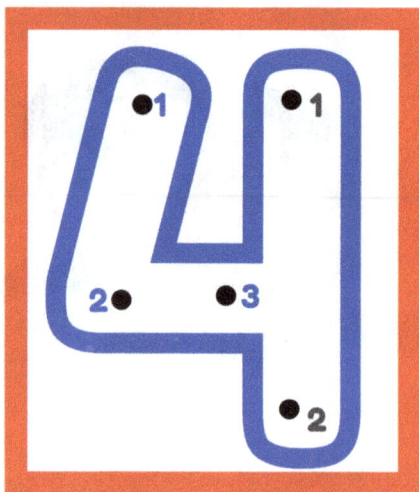

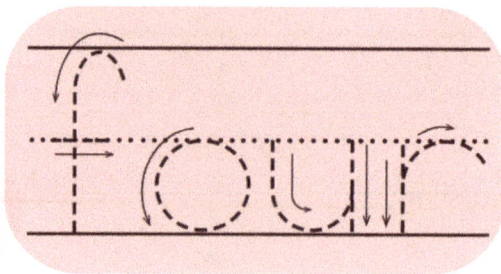

## Trace the Word

four four four

## Write the Word

## Color all the 4's

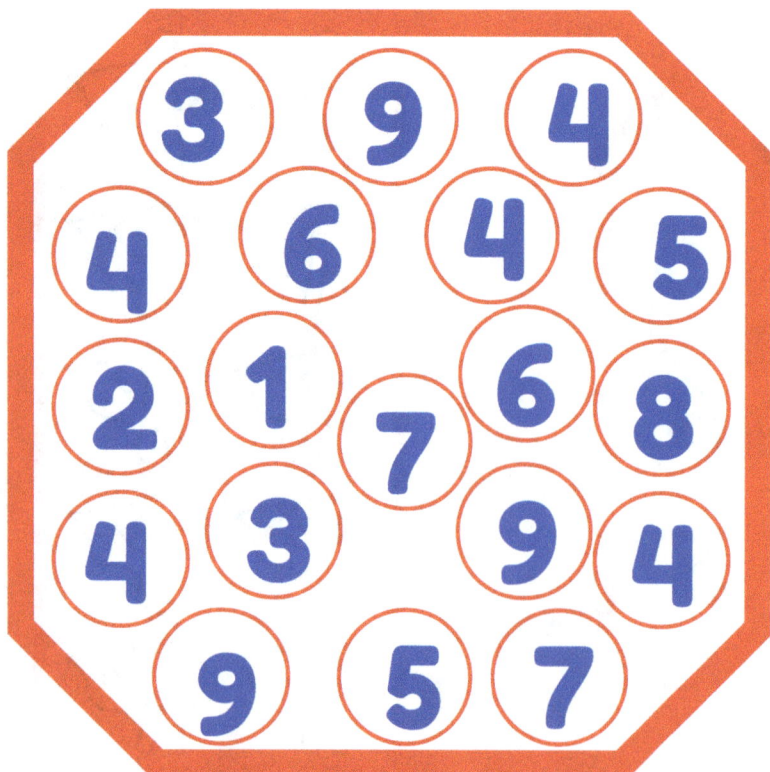

## Color 4 Capes

## Dot to Dot

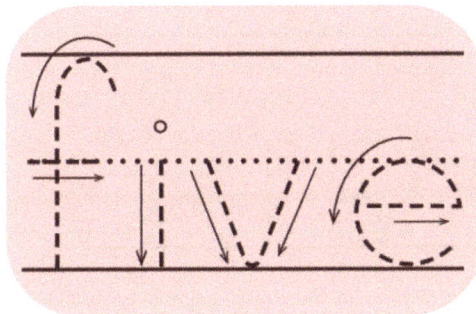

## Trace the Word

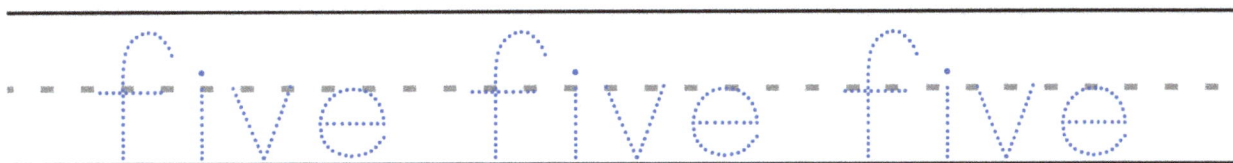

## Write the Word

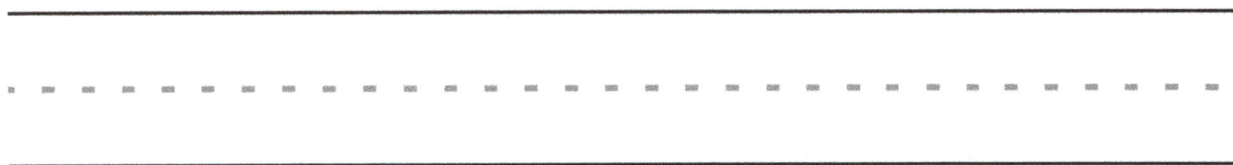

## Color all the 5's

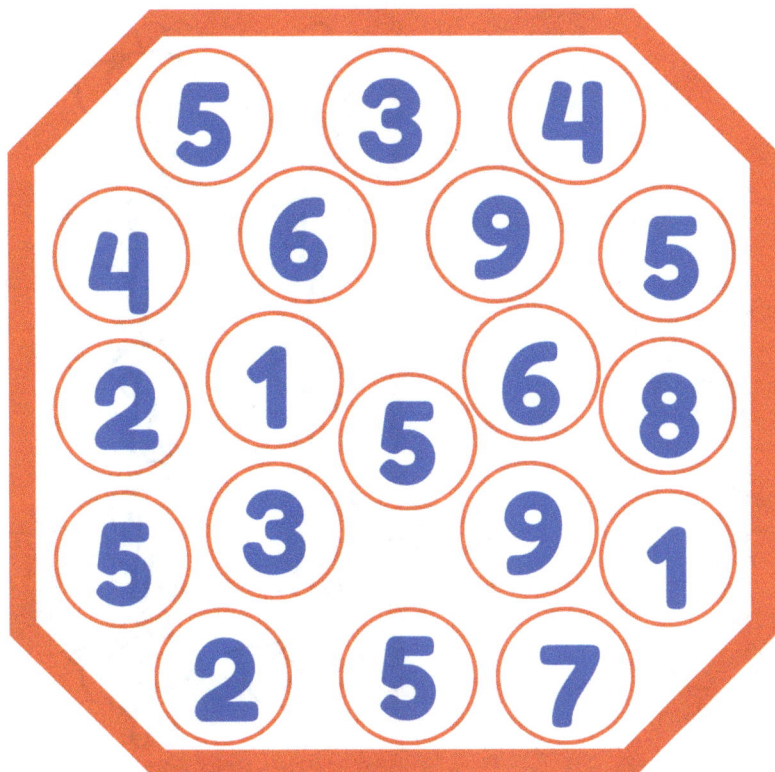

## Color 5 Capes

## Dot to Dot

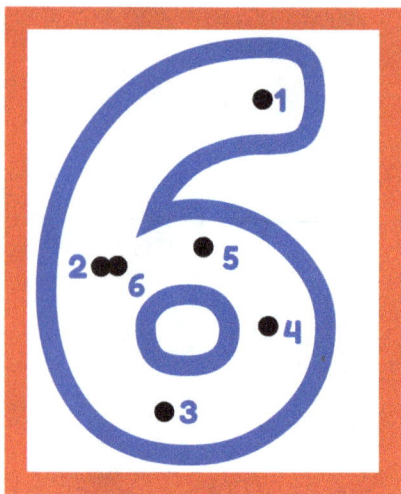

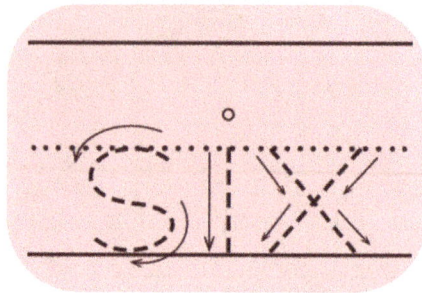

## Trace the Word

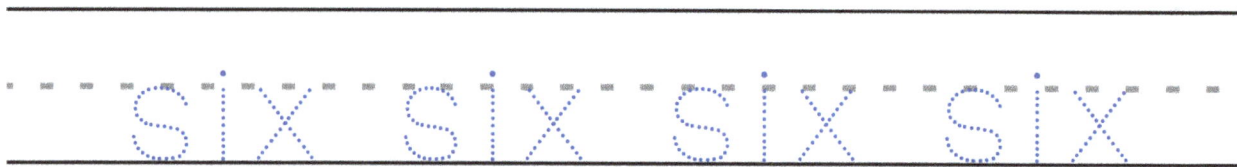

## Write the Word

## Color all the 6's

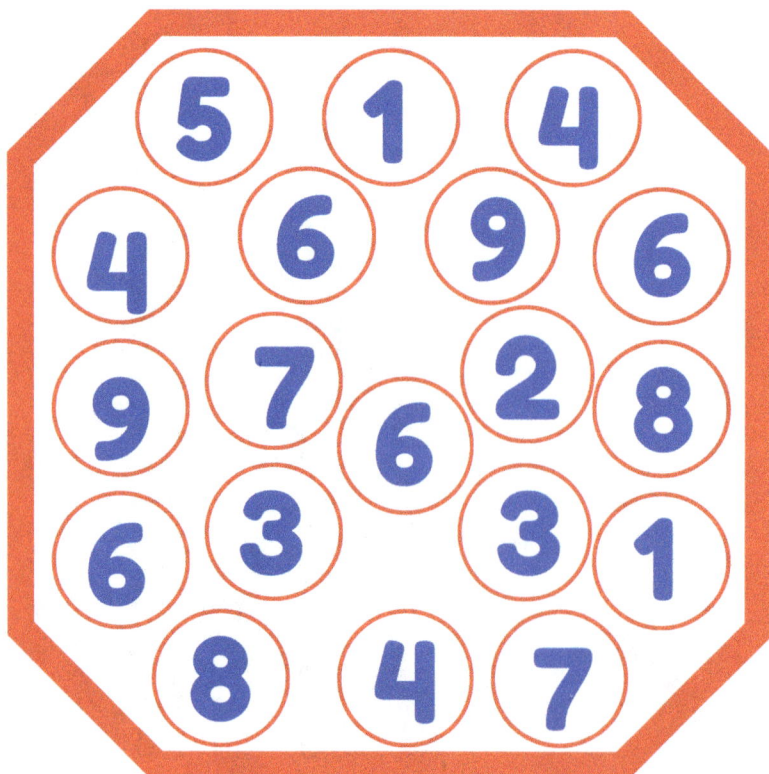

## Color 6 Capes

## Dot to Dot

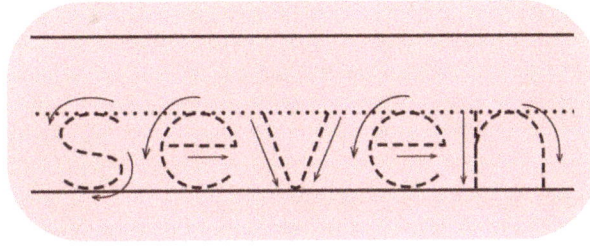

seven

MINTR

## Trace the Word

seven seven seven

## Write the Word

## Color all the 7's

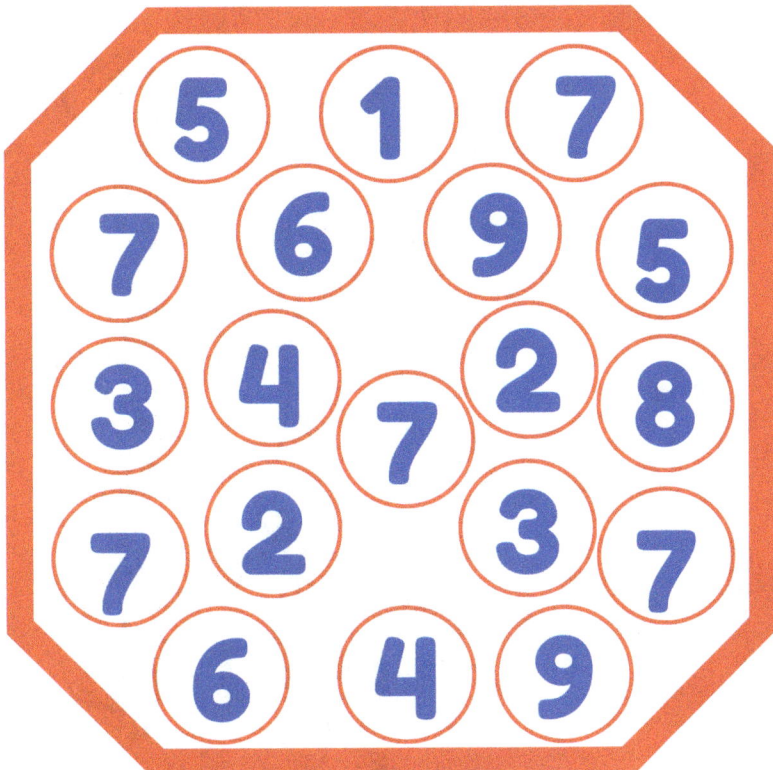

## Color 7 Capes

## Dot to Dot

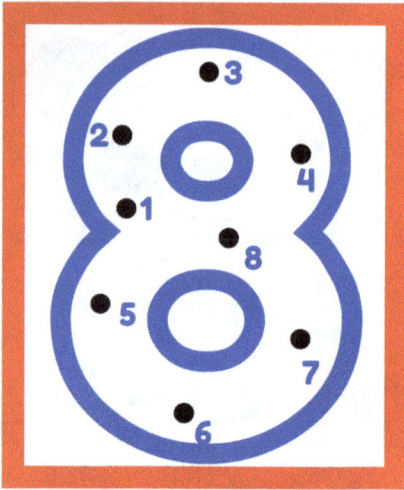

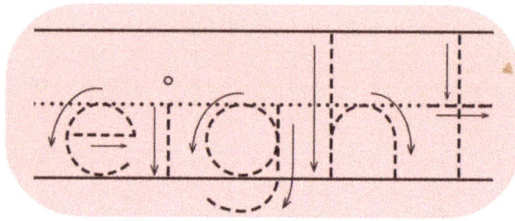

SILVERARROW

## Trace the Word

## Write the Word

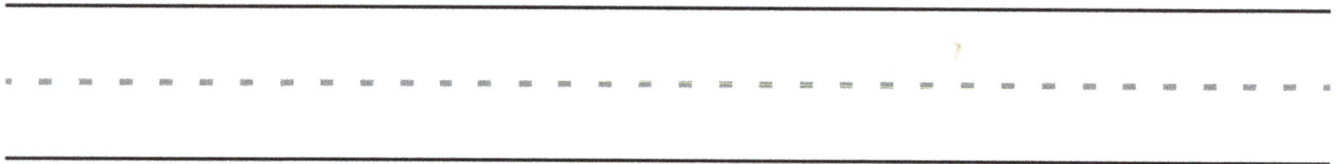

## Color all the 8's

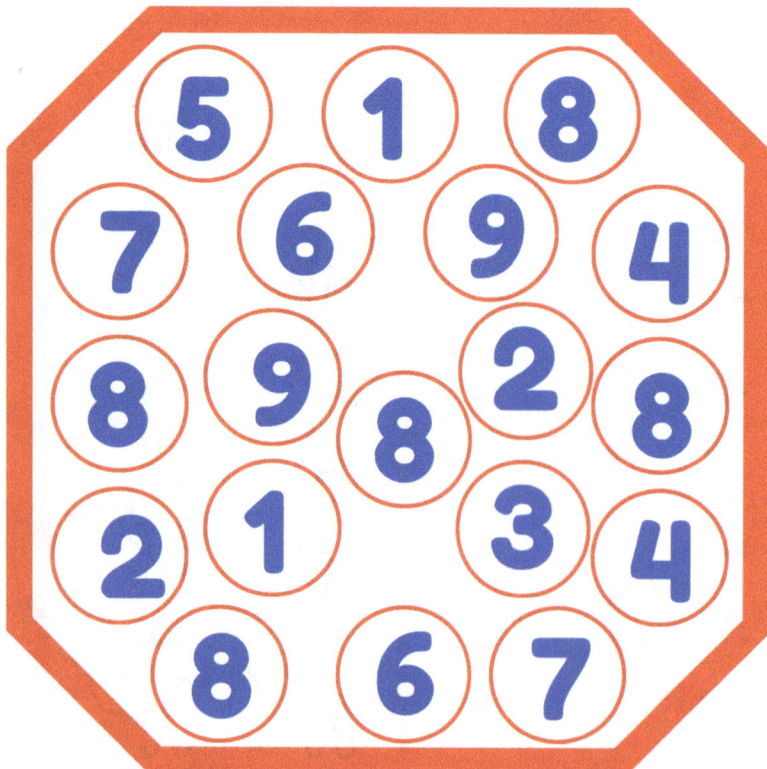

## Color 8 Capes

## Dot to Dot

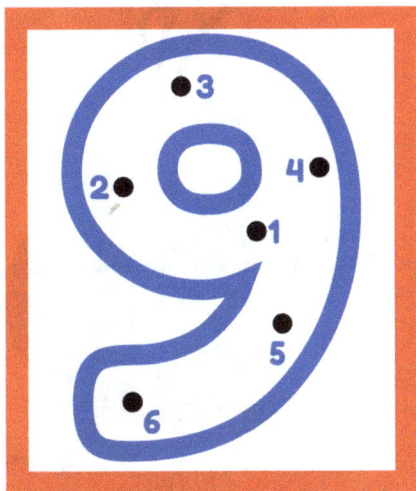

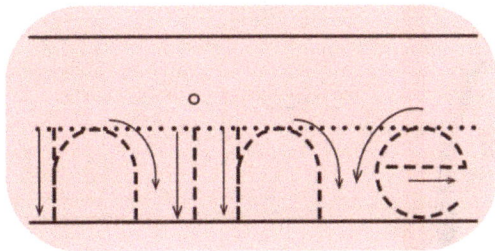

## Trace the Word

nine nine nine

## Write the Word

## Color all the 9's

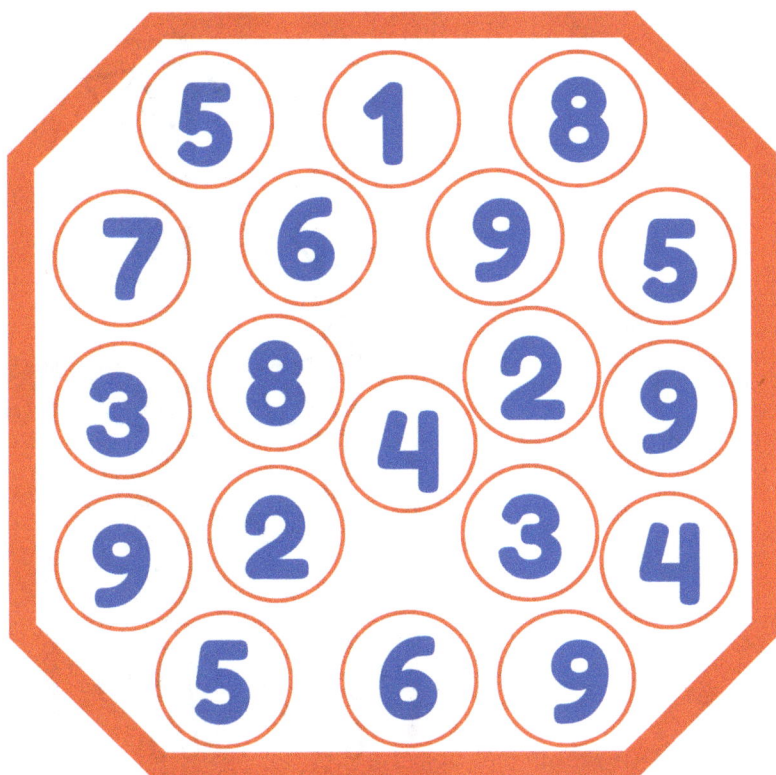

## Color 9 Capes

## Dot to Dot

**10**

**ten**

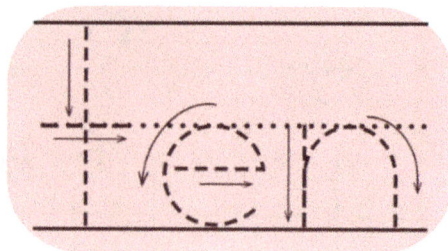

Trace the Word

ten ten ten

Write the Word

## Color all the 10's

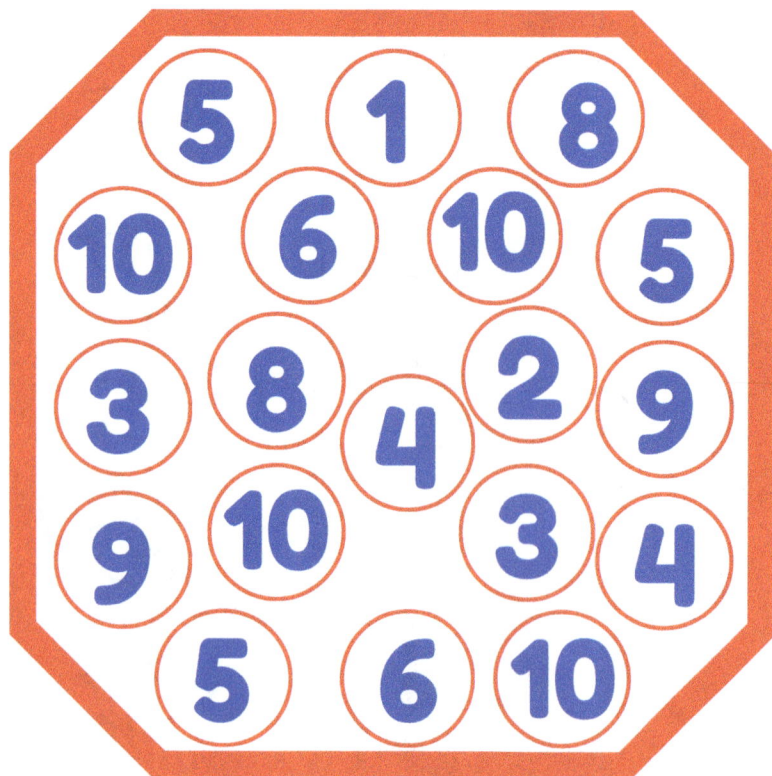

5 1 8
10 6 10 5
3 8 4 2 9
9 10 3 4
5 6 10

## Color 10 Capes

CHRONUS

# YAY!
## You Finished Section 1

Keep Going To Collect All
The Stars!

## Match the Number to the Word.

1. Trace the word.
2. Draw a line to the number.
3. Trace the number.

| | |
|---|---|
| three | 2 |
| six | 7 |
| seven | 3 |
| two | 6 |

**1** Trace the word. Take your time and write the letters carefully.

**2** Find the number that matches the word. Draw a line from the word to the match number.

**3** Trace the number.

# Match the Number to the Word.

1. Trace the word.
2. Draw a line to the number.
3. Trace the number.

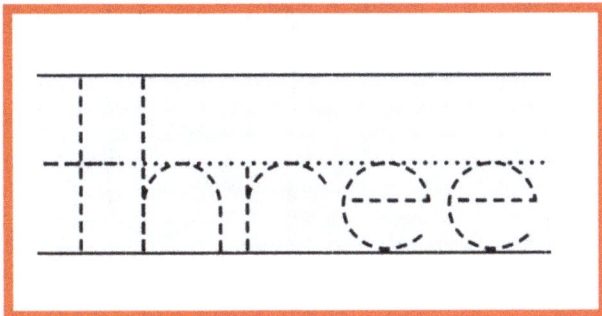

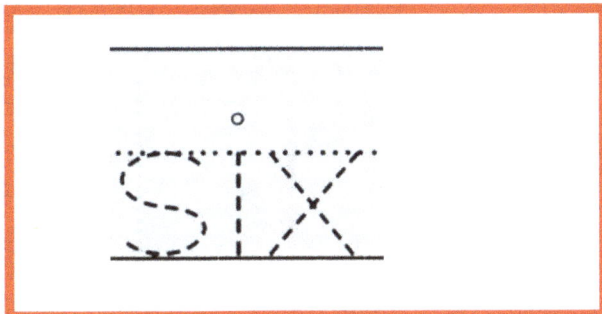

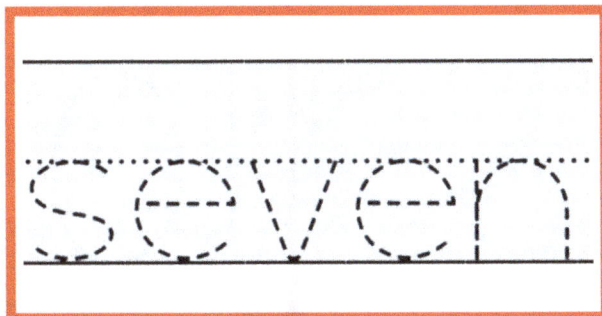

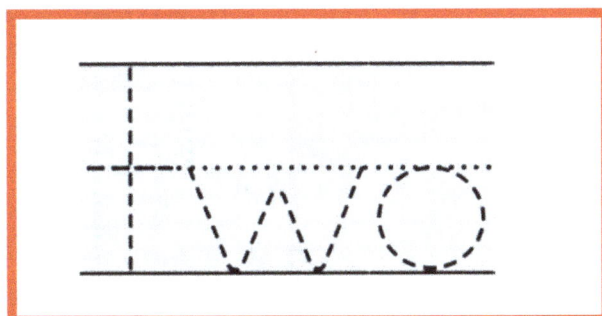

| three | 2 |
| six | 7 |
| seven | 3 |
| two | 6 |

# Match the Number to the Word.

1. Trace the word.
2. Draw a line to the number.
3. Trace the number.

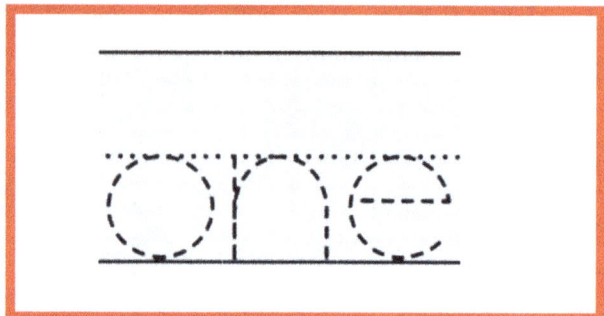

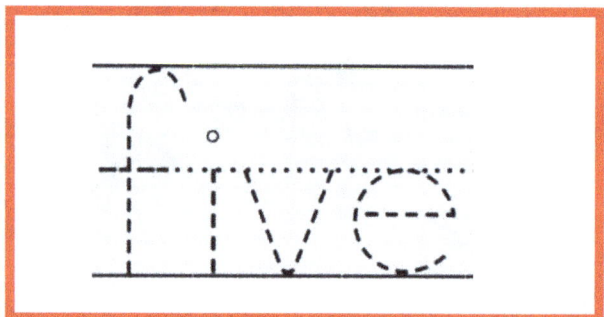

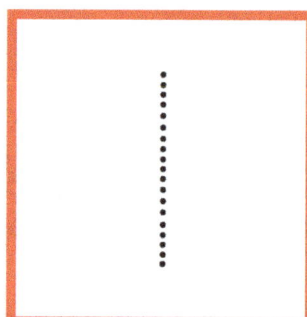

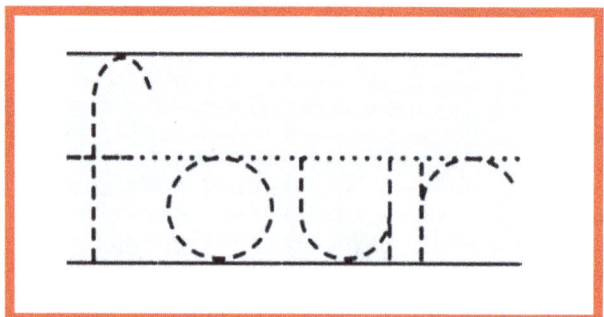

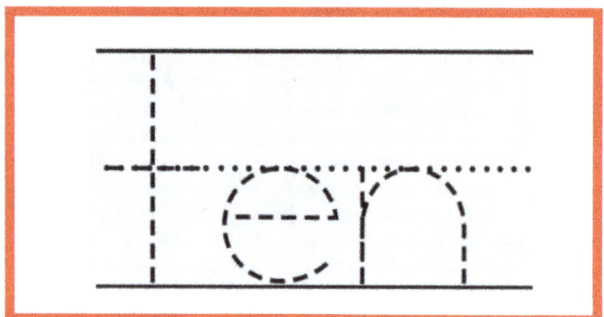

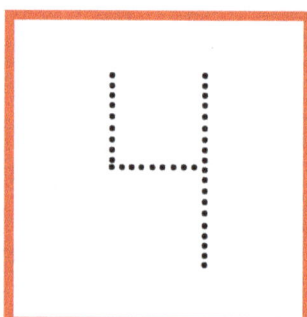

one

five

four

ten

5

1

10

4

# Match the Number to the Word.

1. Trace the word.
2. Draw a line to the number.
3. Trace the number.

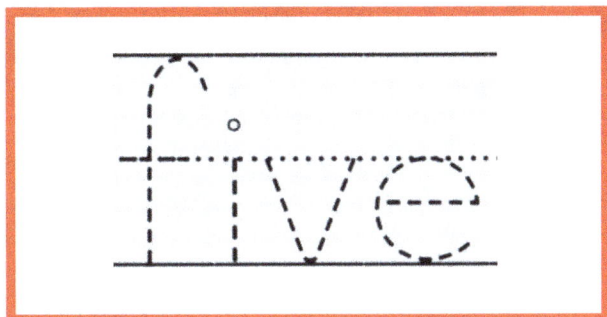

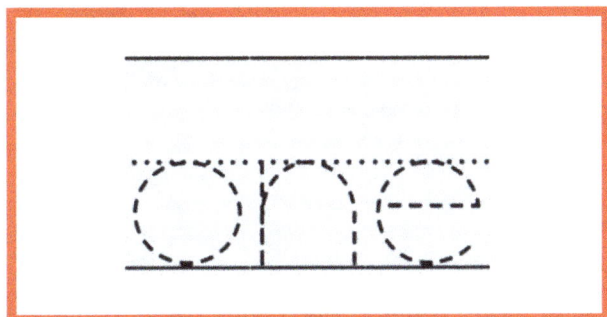

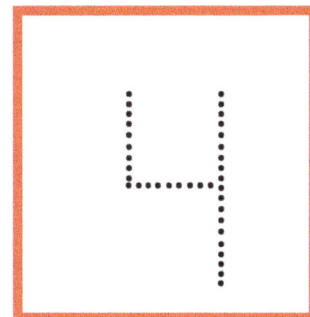

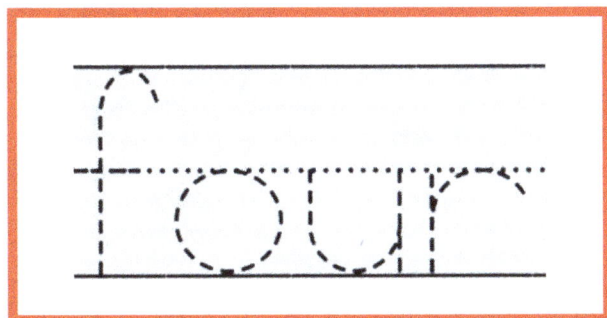

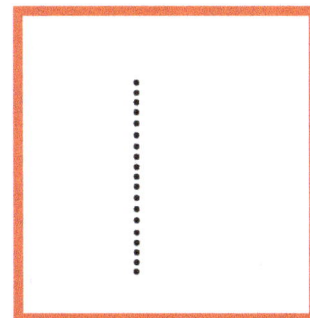

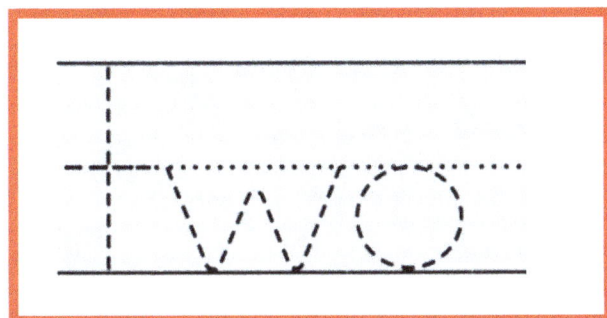

five

one

four

two

2

4

1

5

# Match the Number to the Word.

1. Trace the word.
2. Draw a line to the number.
3. Trace the number.

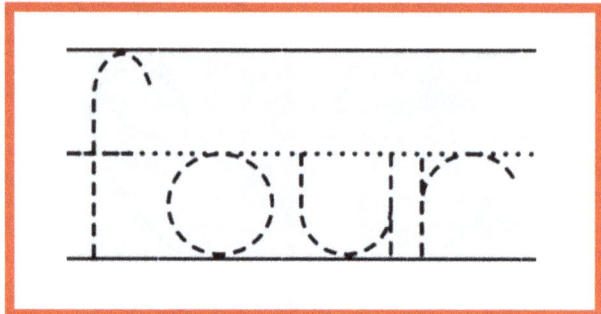

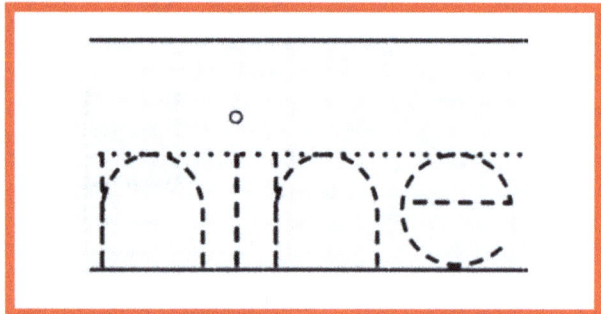

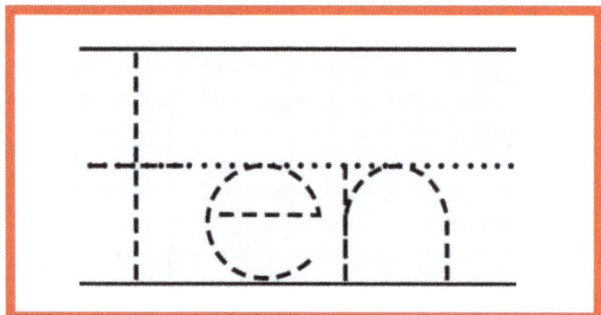

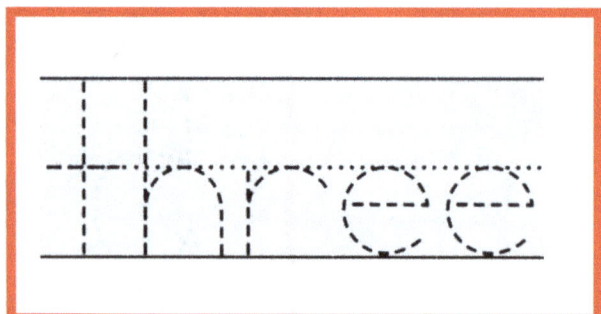

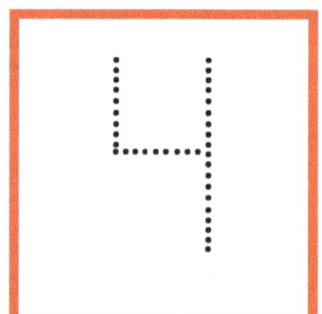

four

nine

ten

three

10

3

9

4

# Match the Number to the Word.

1. Trace the word.
2. Draw a line to the number.
3. Trace the number.

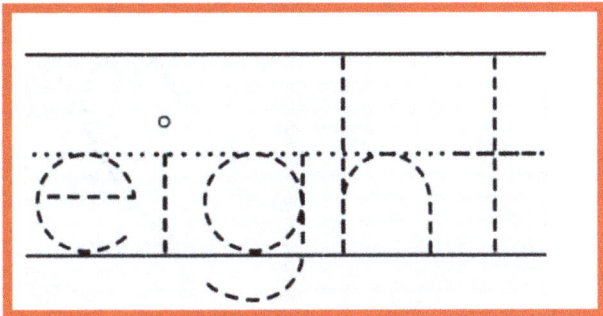

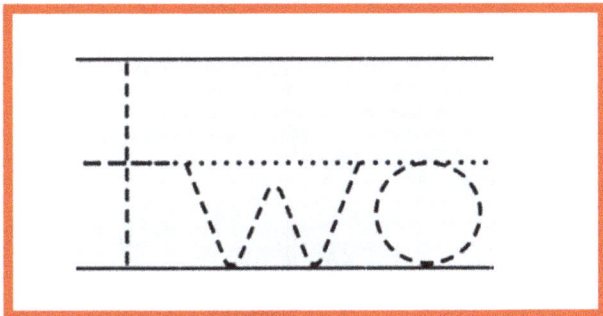

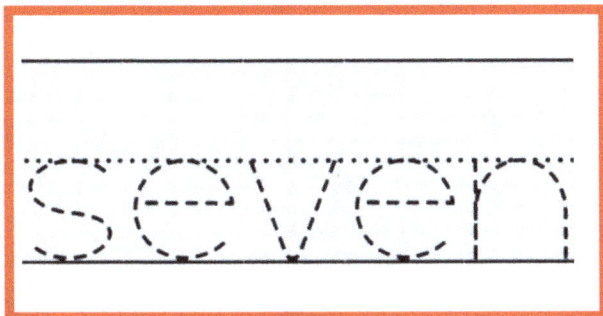

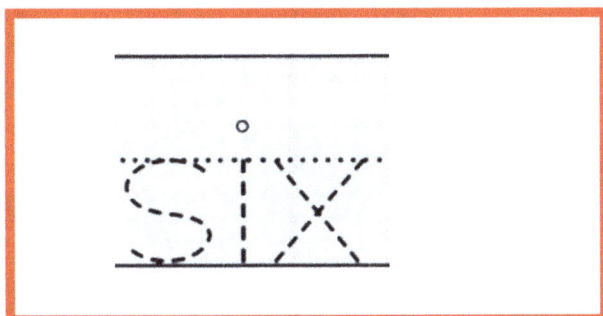

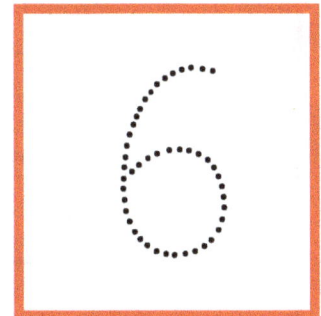

| eight | 7 |
| two | 8 |
| seven | 6 |
| six | 2 |

# Match the Number to the Word.

1. Trace the word.
2. Draw a line to the number.
3. Trace the number.

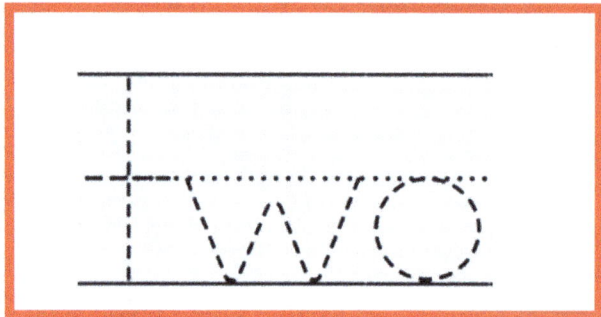

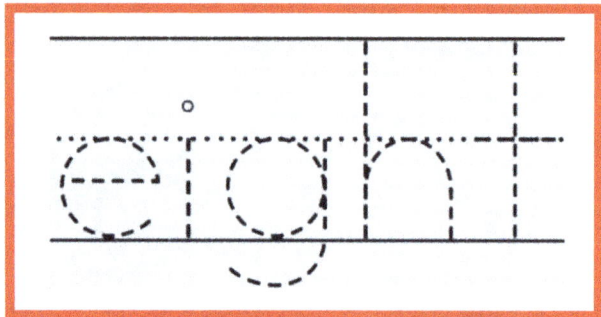

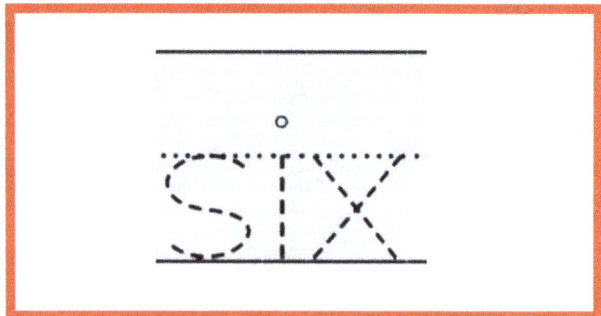

two

eight

six

seven

2

6

8

7

# Match the Number to the Word.

1. Trace the word.
2. Draw a line to the number.
3. Trace the number.

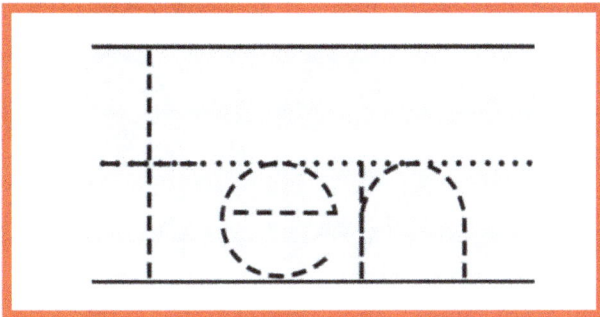

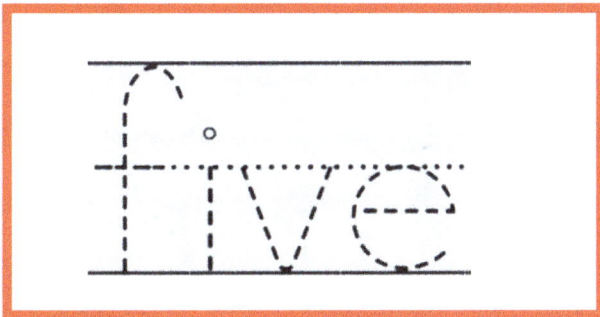

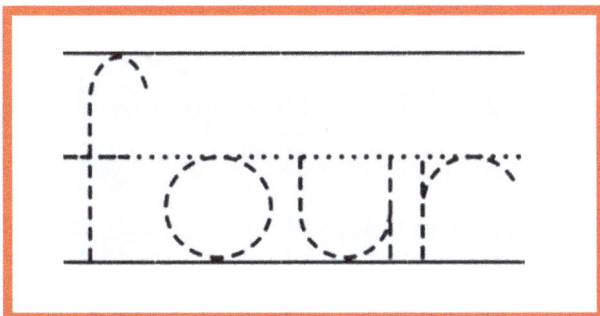

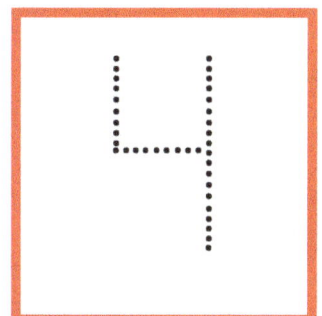

ten

five

four

one

10

1

5

4

# Match the Number to the Word.

1. Trace the word.
2. Draw a line to the number.
3. Trace the number.

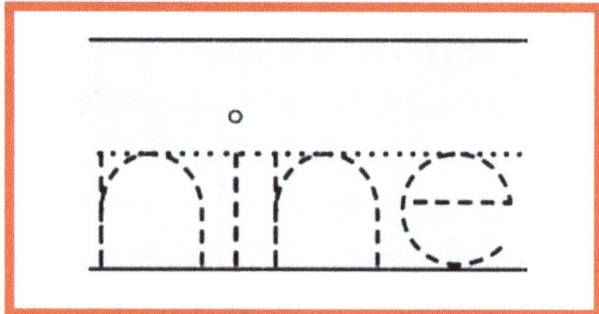

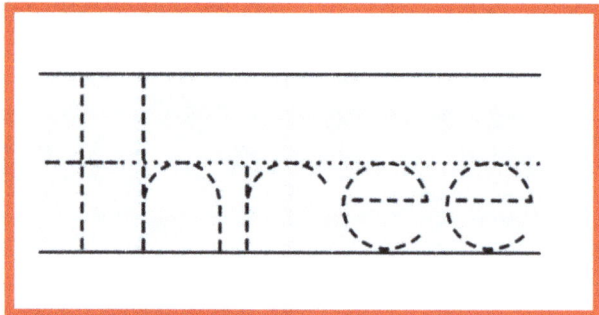

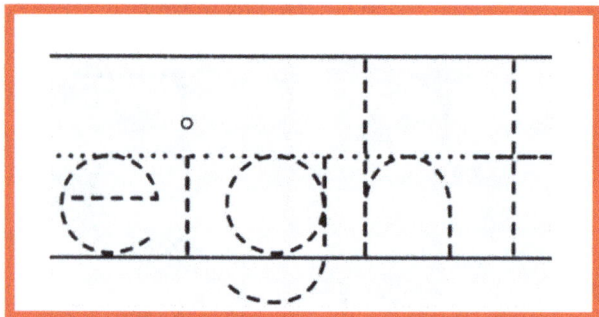

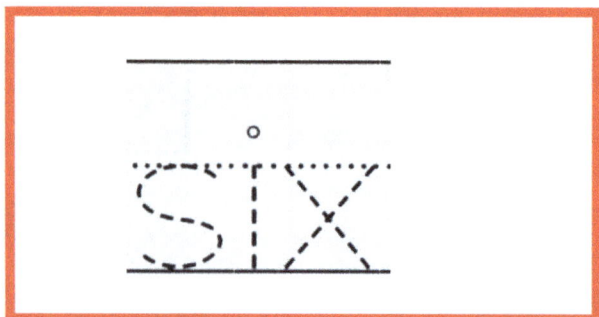

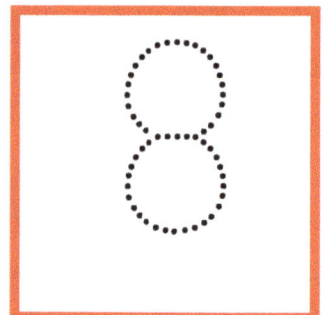

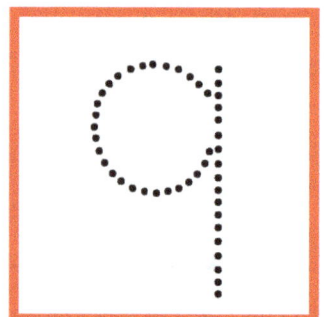

nine

three

eight

six

6

8

9

3

# Match the Number to the Word.

1. Trace the word.
2. Draw a line to the number.
3. Trace the number.

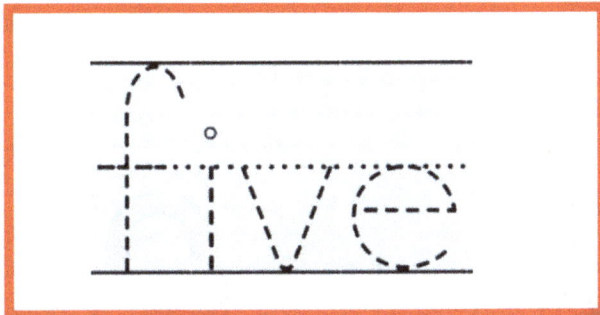

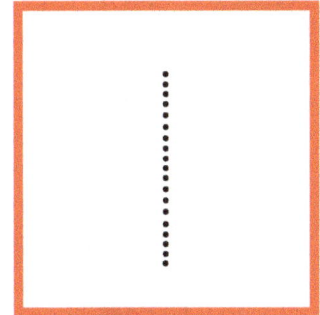

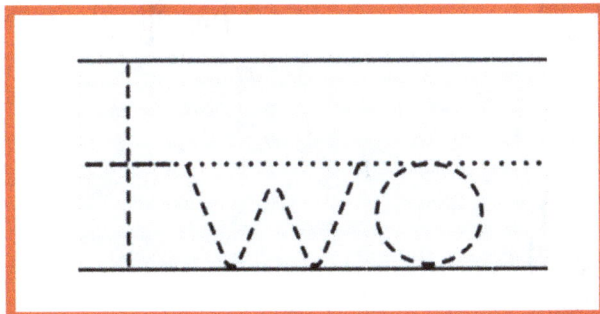

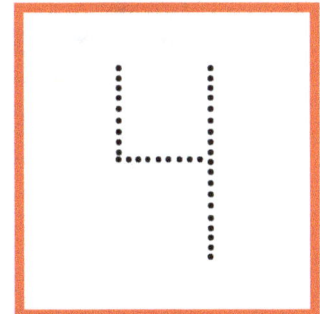

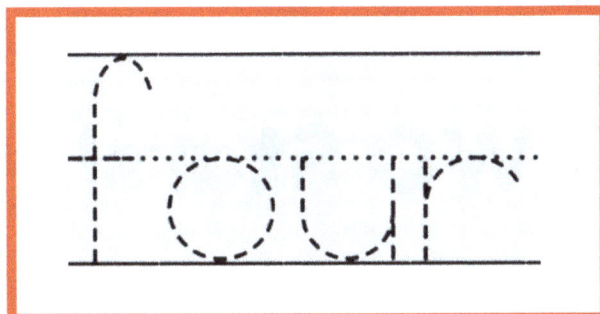

one

five

two

four

2

1

4

5

# WOW!
## Section 2 Finished

You Have TWO Stars!

# Section 3

## What to do.

### Match the Object to the Word.

1. Trace the word.
2. Find the box with that many objects in it.
3. Draw a line to match.

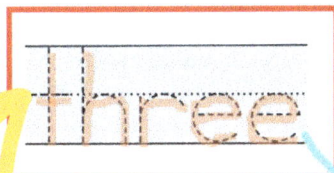

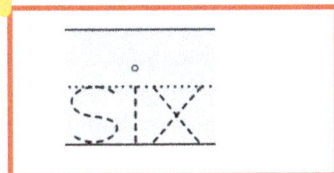

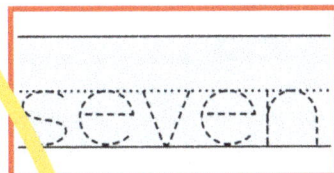

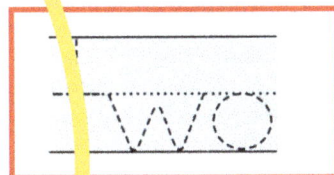

three

six

seven

two

**1** Trace the word. Just like in the last section.

**2** Find the box with that matching number of sweets in it.

**3** Draw a line from the word box to the matching sweets box.

# Match the Object to the Word.

1. Trace the word.
2. Find the box with that many objects in it.
3. Draw a line to match.

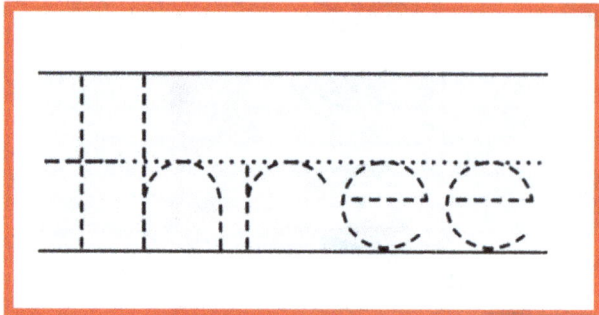

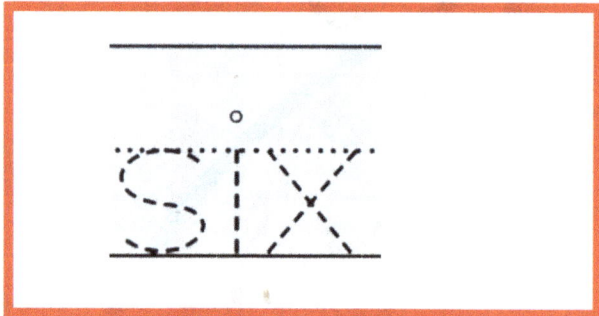

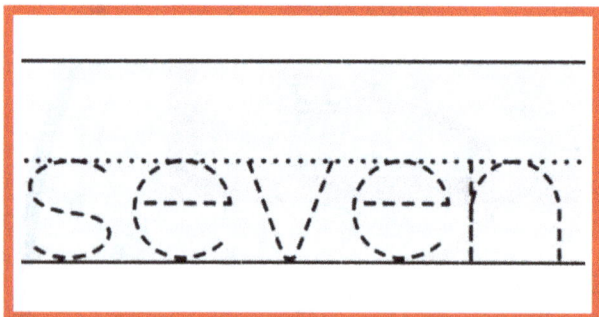

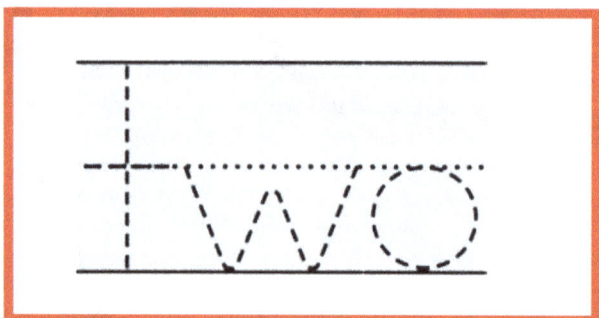

three

six

seven

two

# Match the Object to the Word.

1. Trace the word.
2. Find the box with that many objects in it.
3. Draw a line to match.

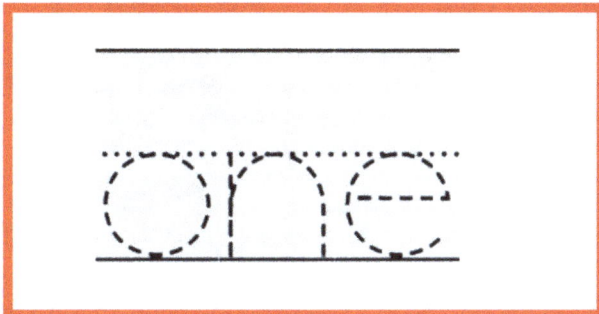

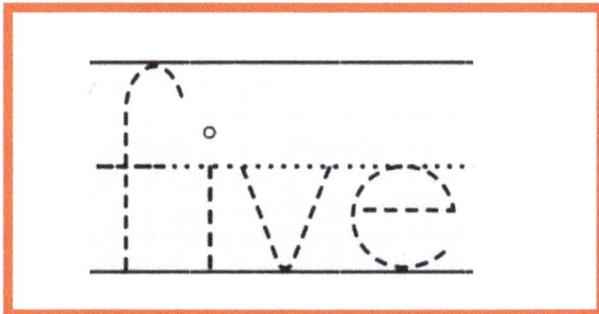

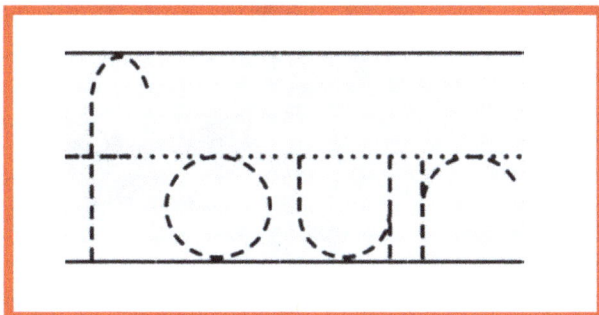

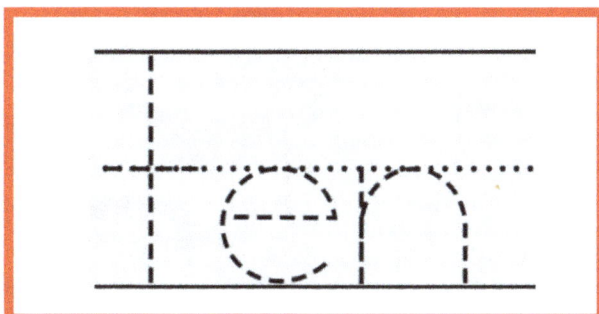

one

five

four

ten

# Match the Object to the Word.

1. Trace the word.
2. Find the box with that many objects in it.
3. Draw a line to match.

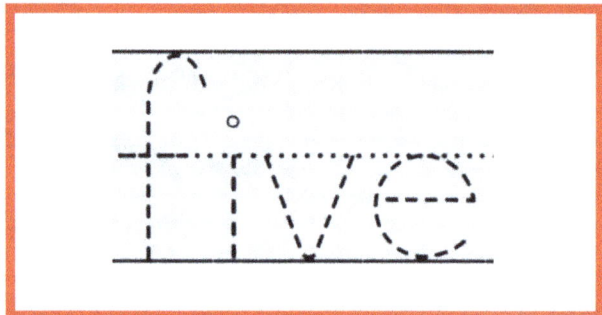

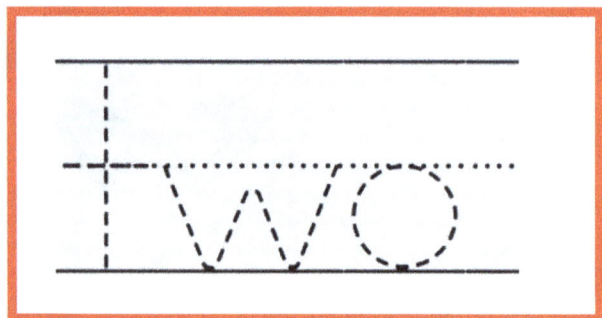

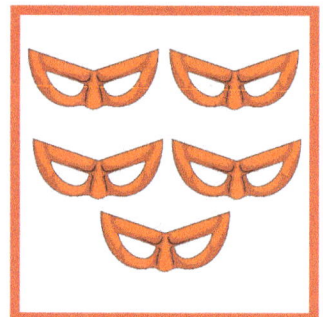

five

one

four

two

# Match the Object to the Word.

1. Trace the word.
2. Find the box with that many objects in it.
3. Draw a line to match.

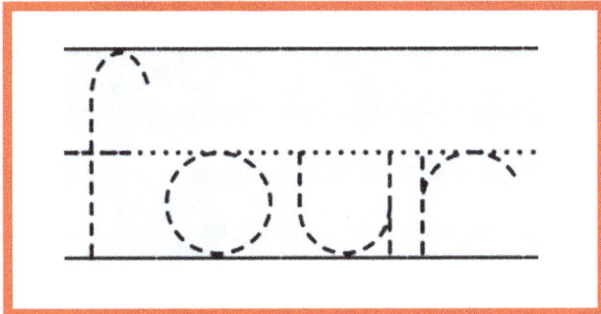

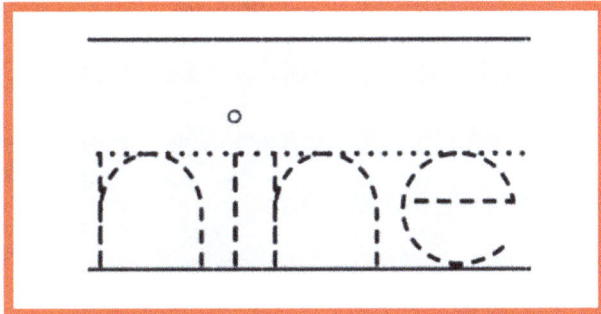

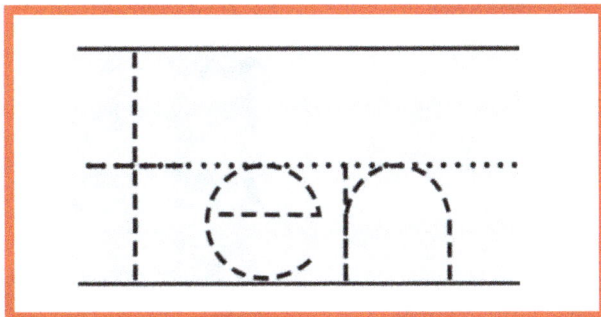

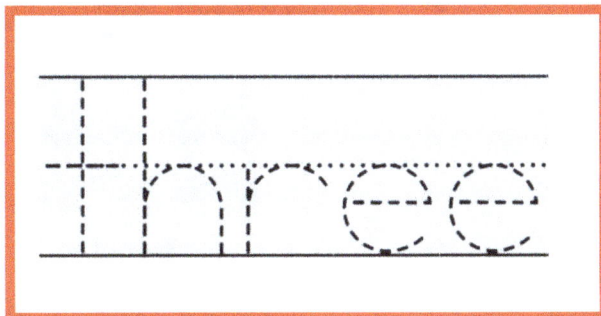

four

nine

ten

three

# Match the Object to the Word.

1. Trace the word.
2. Find the box with that many objects in it.
3. Draw a line to match.

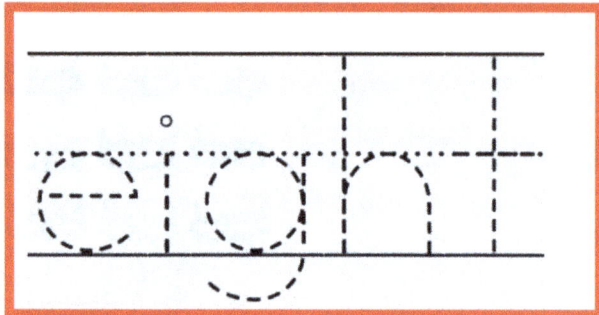

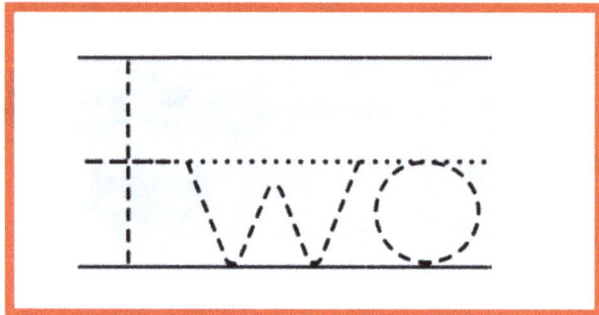

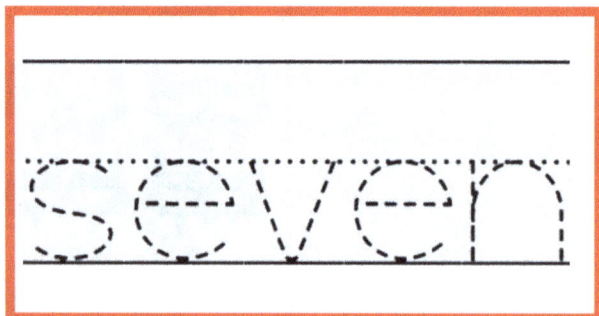

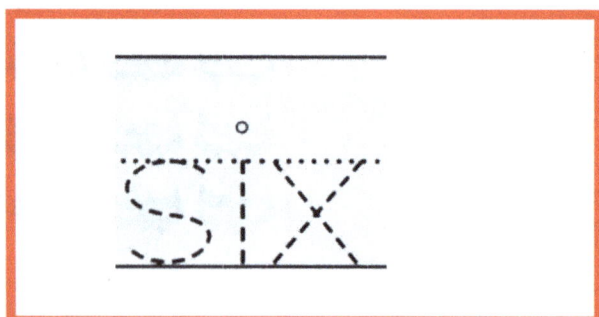

eight

two

seven

six

# Match the Object to the Word.

1. Trace the word.
2. Find the box with that many objects in it.
3. Draw a line to match.

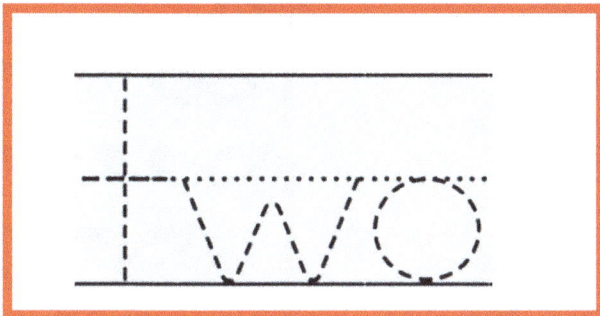

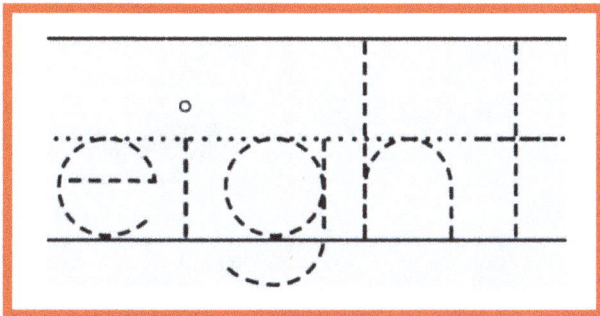

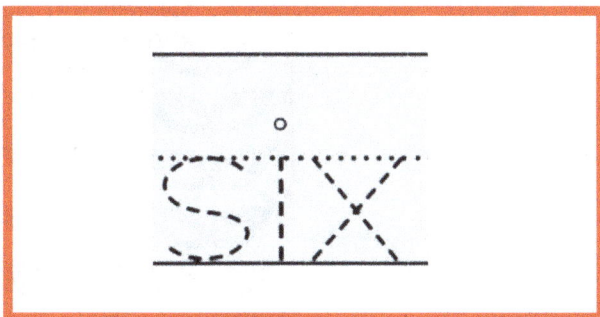

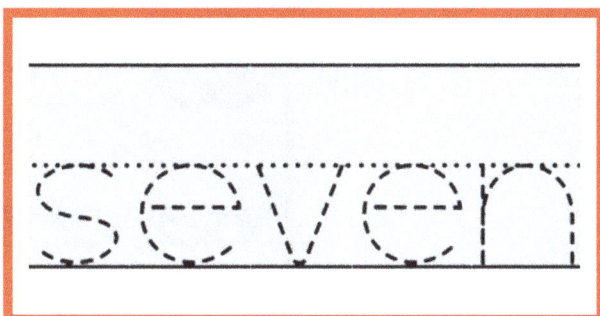

two

eight

six

seven

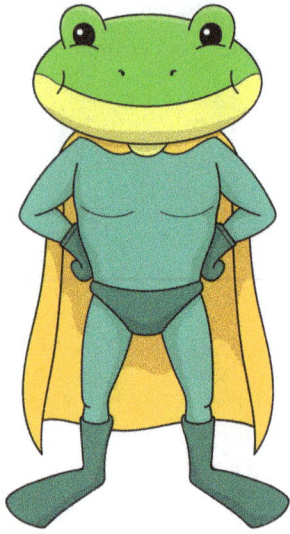

# Match the Object to the Word.

1. Trace the word.
2. Find the box with that many objects in it.
3. Draw a line to match.

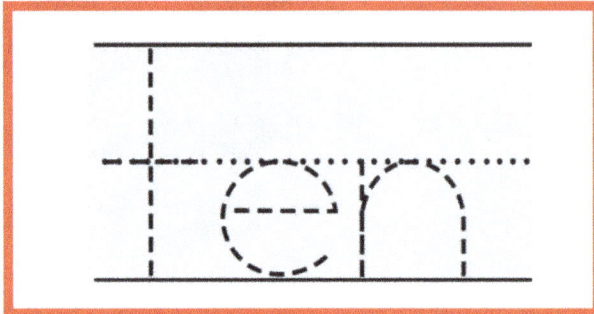

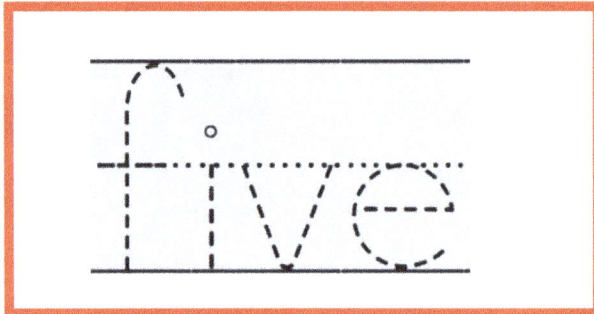

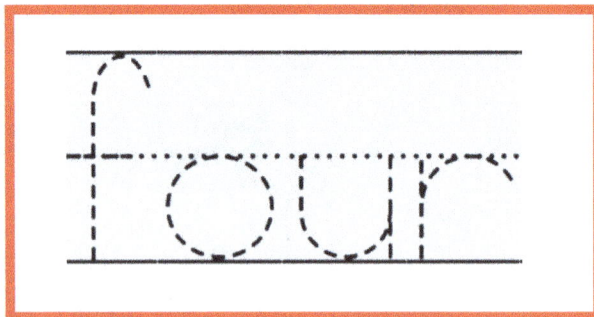

ten

five

four

one

# Match the Object to the Word.

1. Trace the word.
2. Find the box with that many objects in it.
3. Draw a line to match.

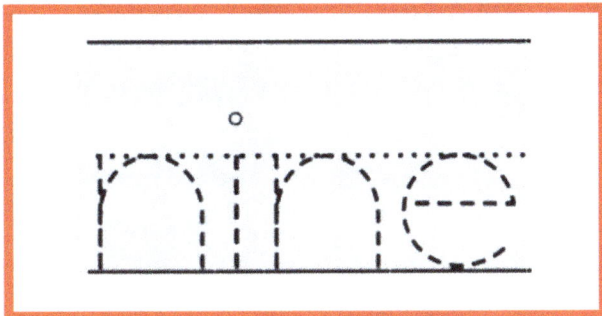

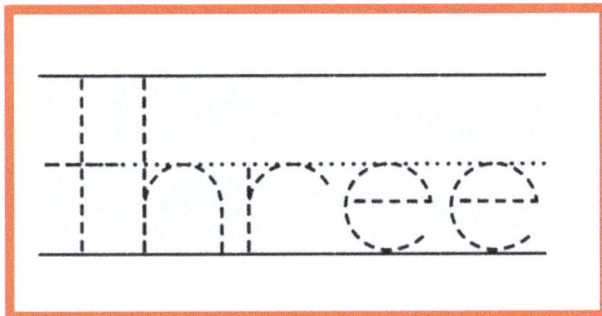

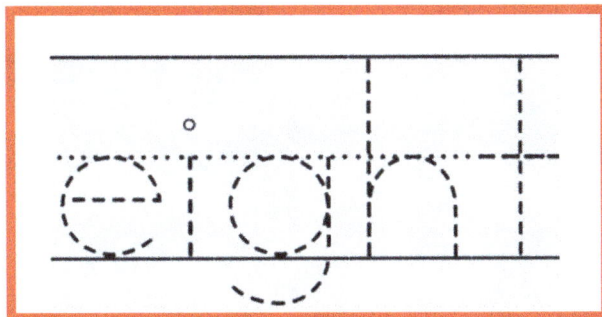

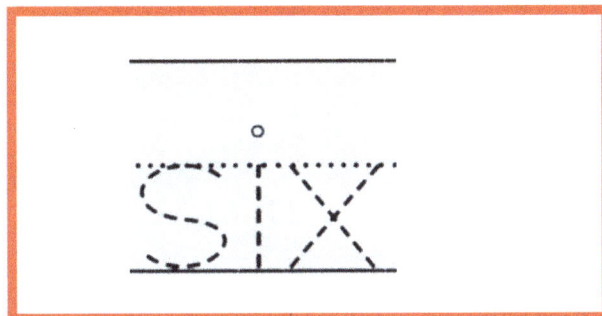

nine

three

eight

six

# Match the Object to the Word.

1. Trace the word.
2. Find the box with that many objects in it.
3. Draw a line to match.

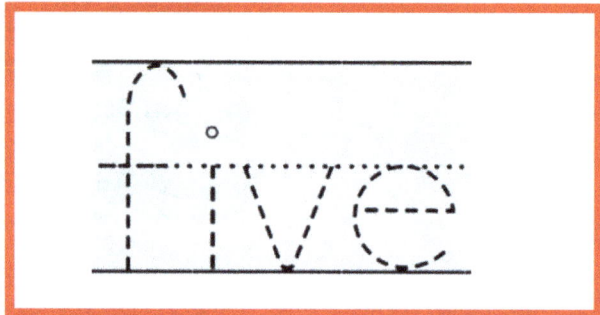

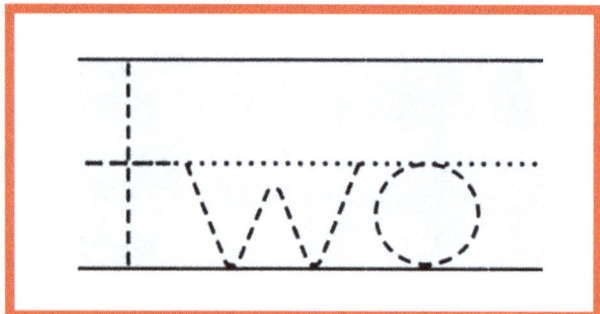

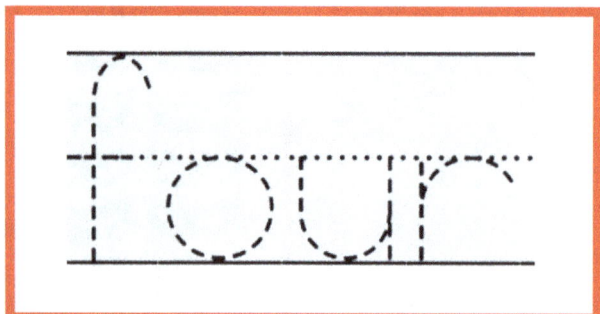

one

five

two

four

# Great Job!
## You Finished Section 3

Keep Going To Collect All The Stars!

## Matching Numbers to Objects

- Draw a line from the picture to the matching number.

| 4 |
| 2 |
| 3 |
| 5 |

**Find the boxes with the same number and the number of treats and draw a line.**

# Matching Numbers to Objects

- Draw a line from the picture to the matching number.

| | |
|---|---|
| (3 shoes) | 4 |
| (5 shoes) | 2 |
| (4 shoes) | 3 |
| (2 shoes) | 5 |

# Matching Numbers to Objects

- Draw a line from the Number to the matching Picture.

| 2 |  |
| 9 |  |
| 7 |  |
| 4 |  |

# Matching Numbers to Objects

- Draw a line from the Number to the matching Picture.

| | |
|---|---|
| **1** |  |
| **6** |  |
| **3** |  |
| **9** |  |

# Matching Numbers to Objects

- Draw a line from the picture to the matching number.

| | |
|---|---|
|  | **6** |
|  | **8** |
|  | **7** |
|  | **4** |

# Matching Numbers to Objects

- Draw a line from the picture to the matching number.

**8**

**6**

**4**

**1**

# Matching Numbers to Objects

- Draw a line from the Number to the matching Picture.

8

4

6

2

# Matching Numbers to Objects

- Draw a line from the Number to the matching Picture.

2

5

9

1

# Matching Numbers to Objects

- Draw a line from the Number to the matching Picture.

3

4

2

8

# Matching Numbers to Objects

- Draw a line from the picture to the matching number.

| | |
|---|---|
|  | **1** |
|  | **7** |
|  | **4** |
|  | **2** |

# Matching Numbers to Objects

- Draw a line from the picture to the matching number.

| | |
|---|---|
| 6 shoes | 4 |
| 4 shoes | 7 |
| 5 shoes | 3 |
| 3 shoes | 6 |

# FANTASTIC!

## Section 4 Finished!

You are over half way done collecting all the Stars!

# ›Section 5‹

In this section, you will practice writing a NUMBER SENTENCE.

A Number Sentence is a Group of Numbers and Symbols that tells us about math problems. Just like how we use words to make sentences to Tell A Story or give information, we use numbers to make number sentences to help us Solve Problems.

## Trace the Numbers

### Addition Facts 1 - 4

You can write a Addition Number Sentence like this.

$1 + 2 = 3$

Trace the numbers

$2 + 2 = 4$

$2 + 1 = 3$

$1 + 3 = 4$

$1 + 1 = 2$

**1** Count how many are in each picture.

**2** Trace the Number on the line below.

# Trace the Numbers

## Addition Facts 1 - 4

You can write a **Addition Number Sentence** like this.

$$1 + 2 = 3$$

**Trace the numbers**

$$2 + 2 = 4$$

$$2 + 1 = 3$$

$$1 + 3 = 4$$

$$1 + 1 = 2$$

# Trace the Numbers

## Addition Number Sentence

## Addition Facts 1 - 4

- Trace each of these numbers in the number sentence.

**Trace the numbers**

1 + 3 = 4

2 + 1 = 3

1 + 1 = 2

2 + 2 = 4

# Trace the Numbers

## Addition Number Sentence

## Addition Facts 1 - 7

- Trace each of these numbers in the number sentence.

Trace the numbers

2 + 3 = 5

3 + 3 = 6

5 + 2 = 7

4 + 1 = 5

# Trace the Numbers

## Addition Number Sentence

## Addition Facts 1 - 7

- Trace each of these numbers in the number sentence.

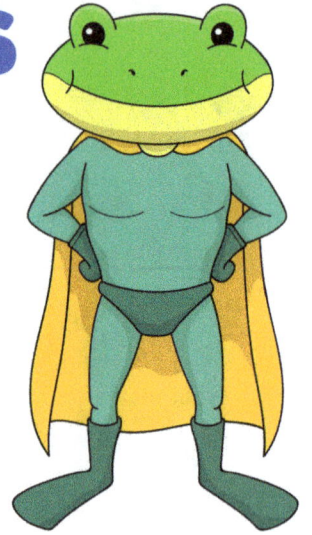

Trace the numbers

4 + 3 = 7

4 + 2 = 6

3 + 3 = 6

2 + 5 = 7

# Trace the Numbers

## Addition Facts 1 - 7

- Trace each of these numbers in the number sentence.

### Trace the numbers

$$4 + 1 = 5$$

$$2 + 3 = 5$$

$$2 + 4 = 6$$

$$5 + 1 = 6$$

# Trace the Numbers

## Addition Number Sentence

## Addition Facts 1 - 10

- Trace each of these numbers in the number sentence.

Trace the numbers

6 + 2 = 8

4 + 5 = 9

4 + 4 = 8

7 + 2 = 9

# Trace the Numbers

## Addition Number Sentence

## Addition Facts 1 - 10

- Trace each of these numbers in the number sentence.

Trace the numbers

5 + 5 = 10

6 + 4 = 10

3 + 7 = 10

1 + 6 = 7

# AWESOME!
## Your So Close.

## Keep Going To Collect All The Stars!

# Missing Number
## Addition Facts 1 - 4

You can write a **Addition Number Sentence** like this.

$$1 + 2 = 3$$

**Fill in the missing numbers.**

$$\underline{\phantom{0}} + 2 = 5$$

$$\underline{\phantom{0}} + 1 = 4$$

$$\underline{\phantom{0}} + 3 = 5$$

$$\underline{\phantom{0}} + 1 = 5$$

**1** Count how many **Cupcakes** there are above the Blank Line

**2** Write that Number on the Blank Line.

# Missing Number

## Addition Facts 1 - 4

You can write a **Addition Number Sentence** like this.

$$1 + 2 = 3$$

**Fill in the missing numbers.**

$$\_\_ + 2 = 5$$

$$\_\_ + 1 = 4$$

$$\_\_ + 3 = 5$$

$$\_\_ + 1 = 5$$

MINTR

# Missing Number

## Addition Facts 1 - 4

- Count how many items you see above the empty line and write that number on the line.

Fill in the missing numbers.

____  +  1  =  5

____  +  3  =  5

____  +  3  =  4

____  +  2  =  4

# Missing Number

## Addition Facts 1 - 7

- Count how many items you see above the empty line and write that number on the line.

Fill in the missing numbers.

_____ + 3 = 4

_____ + 5 = 7

_____ + 1 = 6

_____ + 4 = 5

# Missing Number

## Addition Facts 1 - 7

- Count how many items you see above the empty line and write that number on the line.

**Fill in the missing numbers.**

$$6 + \underline{\phantom{0}} = 7$$

$$4 + \underline{\phantom{0}} = 6$$

$$3 + \underline{\phantom{0}} = 7$$

$$4 + \underline{\phantom{0}} = 5$$

# Missing Number

## Addition Facts 1 - 4

- Count how many items you see above the empty line and write that number on the line.

Fill in the missing numbers.

4 + 1 = ___

3 + 1 = ___

1 + 2 = ___

2 + 2 = ___

# Missing Number

## Addition Facts 1 - 7

- Count how many items you see above the empty line and write that number on the line.

KIMBERLEY

**Fill in the missing numbers.**

4 + 3 = ___

2 + 5 = ___

1 + 4 = ___

3 + 3 = ___

# Missing Number

## Addition Facts 1 - 10

- Count how many items you see above the empty line and write that number on the line.

**Fill in the missing numbers.**

9 + ___ = 10

7 + ___ = 10

5 + ___ = 9

6 + ___ = 8

# Missing Number

## Addition Facts 1 - 10

- Count how many items you see above the empty line and write that number on the line.

**Fill in the missing numbers.**

$$\underline{\quad} + \underline{2} = \underline{7}$$

$$\underline{\quad} + \underline{4} = \underline{8}$$

$$\underline{\quad} + \underline{5} = \underline{9}$$

$$\underline{\quad} + \underline{4} = \underline{9}$$

DUNRAVEN

# Missing Number
## Addition Facts 1 - 10

- Count how many items you see above the empty line and write that number on the line.

SILVERARROW

Fill in the missing numbers.

2 + 4 = ___

4 + 5 = ___

3 + 7 = ___

2 + 5 = ___

# YOU ARE SO CLOSE!

You only need 1 More STAR!

# Section 7

## What to do.

In this section you will put all the parts of a number sentence together like a puzzle, to Finish AND Solve The Number Sentance.

## Number Sentence

### Addition Facts 1 - 4

You can write a Addition Number Sentence like this.

$$1 + 2 = 3$$

Write number sentences about the pictures.

$$1 + 1 = 2$$

$$\_ + \_ = \_$$

$$\_ + \_ = \_$$

$$\_ + \_ = \_$$

Count the number of Ice-cream's and Write it on the Line below to complete each number sentence..

# Number Sentence

## Addition Facts 1 - 4

You can write a **Addition Number Sentence** like this.

1 + 2 = 3

Write number sentences about the pictures.

___ + ___ = ___

___ + ___ = ___

___ + ___ = ___

___ + ___ = ___

# Number Sentence

## Addition Facts 1 - 4

1 + 2 = 3

1. Look at the picture
2. Count how many in each group.
3. Write that number on the line.

Write a number sentence about the pictures like this: 1 + 2 = 3

___ + ___ = ___

___ + ___ = ___

___ + ___ = ___

___ + ___ = ___

# Number Sentence
## Addition Facts 1 - 4

$1 + 2 = 3$

1. Look at the picture
2. Count how many in each group.
3. Write that number on the line.

**Write a number sentence about the pictures like this:**

___  +  ___  =  ___

___  +  ___  =  ___

___  +  ___  =  ___

___  +  ___  =  ___

# Number Sentence

## Addition Facts 1 - 7

🍦 + 🍦🍦 🍦🍦🍦
_1_ _2_ = _3_

1. Look at the picture
2. Count how many in each group.
3. Write that number on the line.

Write a number sentence about the pictures like this:

🍦🍦🍦🍦 **+** 🍦🍦🍦 **=** 🍦🍦🍦🍦🍦🍦🍦

____ ____ ____

🍦🍦🍦 **+** 🍦🍦🍦 **=** 🍦🍦🍦🍦🍦🍦

____ ____ ____

🍦🍦🍦 **+** 🍦🍦 **=** 🍦🍦🍦🍦🍦

____ ____ ____

🍦🍦🍦🍦🍦 **+** 🍦🍦 **=** 🍦🍦🍦🍦🍦🍦🍦

____ ____ ____

# Number Sentence
## Addition Facts 1 - 7

$$1 + 2 = 3$$

1. Look at the picture
2. Count how many in each group.
3. Write that number on the line.

Write number sentences about the pictures.

\_\_\_ + \_\_\_ = \_\_\_

\_\_\_ + \_\_\_ = \_\_\_

\_\_\_ + \_\_\_ = \_\_\_

\_\_\_ + \_\_\_ = \_\_\_

# Number Sentence

## Addition Facts 1 - 7

$$1 + 2 = 3$$

1. Look at the picture
2. Count how many in each group.
3. Write that number on the line.

Write number sentences about the pictures.

3 + 1 = 4
____ + ____ = ____

3 + 3 = 6
____ + ____ = ____

3 + 3 = 6
____ + ____ = ____

4 + 1 = 5
____ + ____ = ____

# Number Sentence
## Addition Facts 1 - 10

$$\underline{1} + \underline{2} = \underline{3}$$

1. Look at the picture
2. Count how many in each group.
3. Write that number on the line.

**Write number sentences about the pictures.**

$$\underline{\phantom{0}} + \underline{\phantom{0}} = \underline{\phantom{0}}$$

$$\underline{\phantom{0}} + \underline{\phantom{0}} = \underline{\phantom{0}}$$

$$\underline{\phantom{0}} + \underline{\phantom{0}} = \underline{\phantom{0}}$$

$$\underline{\phantom{0}} + \underline{\phantom{0}} = \underline{\phantom{0}}$$

# Number Sentence

## Addition Facts 1-10

🍦 + 🍦🍦 🍦🍦🍦
_1_  _2_ = _3_

1. Look at the picture
2. Count how many in each group.
3. Write that number on the line.

Write number sentences about the pictures.

🍦🍦🍦🍦🍦 + 🍦🍦🍦 = 🍦🍦🍦🍦🍦🍦🍦🍦

___ ___ ___

🍦🍦🍦 + 🍦🍦🍦 = 🍦🍦🍦🍦🍦🍦

___ ___ ___

🍦🍦🍦🍦🍦 + 🍦🍦🍦🍦 = 🍦🍦🍦🍦🍦🍦🍦🍦🍦

___ ___ ___

🍦🍦🍦🍦🍦🍦 + 🍦🍦🍦 = 🍦🍦🍦🍦🍦🍦🍦🍦🍦

___ ___ ___

# Number Sentence
## Addition Facts 1-10

1 + 2 = 3

1. Look at the picture
2. Count how many in each group.
3. Write that number on the line.

Write number sentences about the pictures.

___ + ___ = ___

___ + ___ = ___

___ + ___ = ___

___ + ___ = ___

# YOU DID IT!

# YOUR A STAR!

# WANT MORE SUPERHERO FRIENDS?

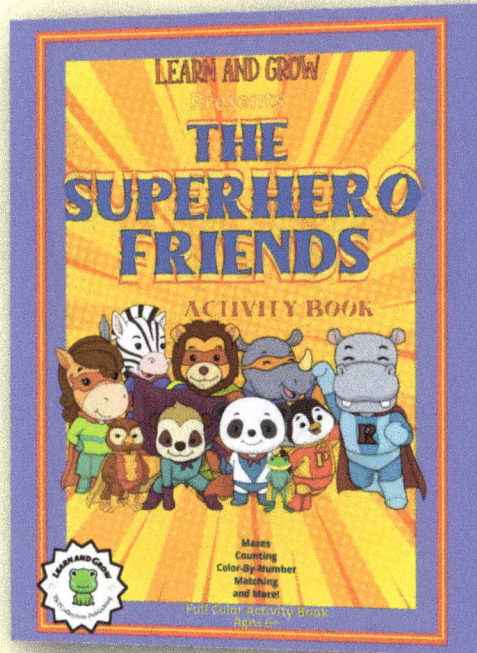

LEARN AND GROW
Presents

# THE SUPERHERO FRIENDS

## ACTIVITY BOOK

Mazes
Counting
Color-By-Number
Matching
and More!

Full Color Activity Book
Ages 6+

## SCAN HERE

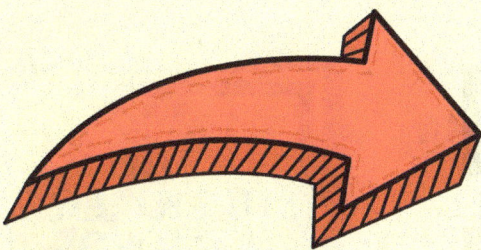

**TKTCollection Publishing: books, biography, latest update**

amazon.com

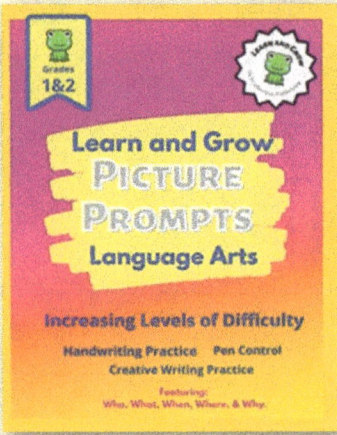

# FIND MORE

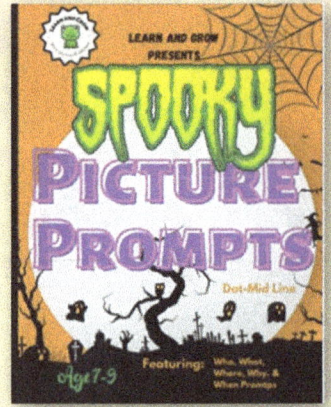

# LEARN AND GROW

# BOOKS

# @

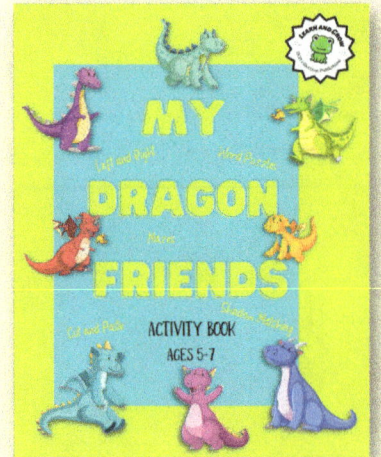

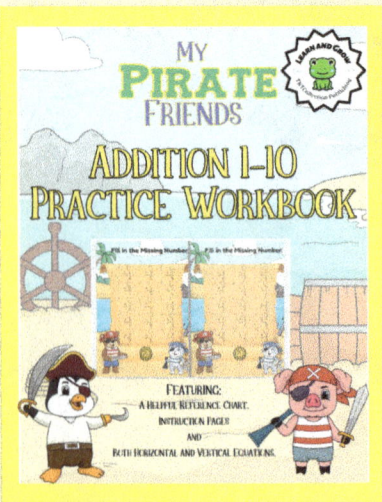

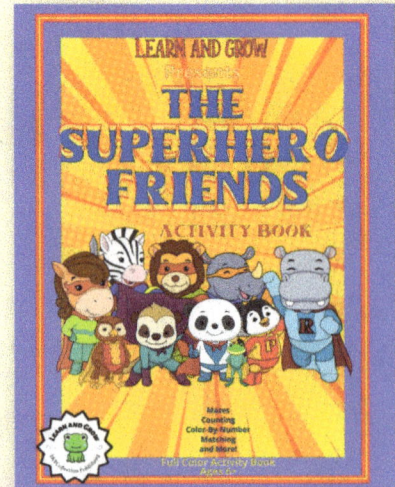

TKTCOLLECTION.COM